Adel Amar Amouri

# Basic concepts in plant breeding and genetic engineering

Adel Amar Amouri

# Basic concepts in plant breeding and genetic engineering

ScienciaScripts

**Imprint**
Any brand names and product names mentioned in this book are subject to trademark, brand or patent protection and are trademarks or registered trademarks of their respective holders. The use of brand names, product names, common names, trade names, product descriptions etc. even without a particular marking in this work is in no way to be construed to mean that such names may be regarded as unrestricted in respect of trademark and brand protection legislation and could thus be used by anyone.

Cover image: www.ingimage.com

This book is a translation from the original published under ISBN 978-613-9-54777-7.

Publisher:
Sciencia Scripts
is a trademark of
Dodo Books Indian Ocean Ltd. and OmniScriptum S.R.L publishing group

120 High Road, East Finchley, London, N2 9ED, United Kingdom
Str. Armeneasca 28/1, office 1, Chisinau MD-2012, Republic of Moldova, Europe
Managing Directors: Ieva Konstantinova, Victoria Ursu
info@omniscriptum.com

Printed at: see last page
**ISBN: 978-620-8-53439-4**

## AVANT - ABOUT US

*This course handout deals with a science that is fairly important for sustainable development and food security in field of agriculture: plant breeding. From a multidisciplinary point of view, its aim is create new elite varieties known as "ideotypes" of plants that meet the nutritional needs of populations. Plant breeding is constantly evolving, especially with the development of recent biotechnological techniques in Molecular Biology and Genetic Engineering, which have contributed effectively to the creation of new plant varieties combining favourable genes coding for genes of interest, based on the great existing plant diversity.*

*This booklet is intended mainly for university students enrolled in the LMD in Biological Sciences, Precision Agriculture and Agronomy who have a basic knowledge of fundamental genetics and who want to learn more about this field. This book can be a teaching tool and a good guide for students involved in basic and professional scientific research.*

***Dr AMOURI Adel Amar***

*Enseignant Chercheur /Maître de Conférence Department of Biology. BP 1524. SNV Faculty. University of Oran 1. 31000-Oran -Algeria amouri.adel@univ-oran1.dz*

## TABLE OF CONTENTS

# 1. WHAT IS PLANT IMPROVEMENT?

## 1.1.Definition of Jules Bouharmont

According to Bouharmont (1994), yield improvements are essential if we are to meet the growing needs of humanity. This improvement can be very rapid for recently domesticated species (oil palm, rubber).

Significant progress has been made with traditional cereals and other plants that have been cultivated for thousands of years. However, there is a limit to the amount of organic matter that can be produced in a given region, depending on temperature and insolation.

One constant objective is to distribute the products of the plant's metabolism in favour of the harvested organs, keeping waste to a minimum: in cereals, this distribution implies maximum development of the grain and a reduced vegetative apparatus.

In addition to productivity, the breeder must also take into account more qualitative characteristics such as nutritional value, organoleptic properties and product composition, aesthetics, resistance, ease of harvesting, transport, storage and processing yield.

In short, the breeder must set himself an ideotype, an ideal image of the plant he wants to create. A very precise ideotype, however, is utopian, because too many often contradictory factors come into play, between which a balance must be found.

On the other hand, creating a new variety generally takes many years, during which time the breeder must be able to modify his objectives and methods to adapt to new conditions, correct his mistakes or benefit from the experience of his competitors.

## 1.2. Definition of Yves Demarly

According to Demarly (1996), hundreds of thousands of years ago, when one of our ancestors brought fruit back to his hut to eat with his family and left seeds in detritus to germinate near his home, plant improvement had begun. First unconscious domestication, then gardening, exchanges, codification of agronomy, control of descendants reinforced by the first genetic laws, then the development of plant improvement as a science and the eruption of biotechnologies. And yet is the creation of a new variety a real science? Drawing the profile of an ideotype that better meets the needs of mankind and its industries is also an artistic and intuitive work, sensitive as much to the comfort of the user as to the technical potential of the material.

On the other hand, it is a science to place the dream object in concrete terms and to calculate the most efficient and economical path from the source material to this ideotype. It is in this compromise between science and art that plant improvement played out (Fig. 1).

Yields on major crops such as wheat, beet, maize, rape, etc. are constantly rising. Statistically, this increase can be put at 1% on average each year.

Of course, a large part of these increases must be attributed to pyrotechnical improvements (fertilisers, treatments, machinery). But estimates from a variety of sources suggest that half of the gains made must be to genetic progress, in the form of the introduction new varieties. Although these ever-increasing yields have sometimes been achieved at the expense of quality, this is only the case in very specific instances.

The standardisation of wheat imposes well-defined baking value standards which have not slowed down production gains, while the dietary value of rapeseed oil has increased at the same time.

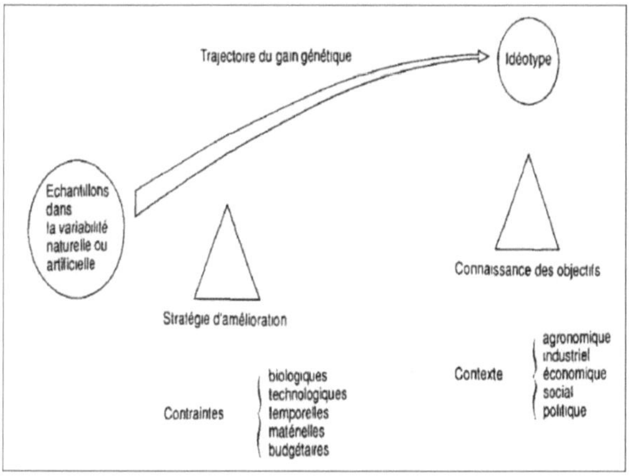


Fig 1: **Definition Plant Improvement** (Demarly, 1996)

### 1.3. Definition by André Gallais

According to Gallais (2015), humans began cultivating plants around 10,000 years ago. As a result of the profound climatic changes that took place at that time, and because his gathering, hunting and fishing activities were no longer sufficient to feed him, he moved from a nomadic to a sedentary state, and began to harvest the products from the plants he had sown.

For the species chosen, the alternation of sowing and harvesting over many cycles then led to their adaptation to cultivation. This is what is known as plant domestication. It is a selection process, based on growing conditions, harvesting conditions and the choices made by man who selected and reseeded the plants best suited to his tastes and needs.

This was the first form of selection made by man to improve his diet, even if it was not always conscious.

Since then, aim of agriculture has always been to produce enough, both in terms of quality and quantity. We need to improve both the quantity and the quality of our produce, in order to feed people better.

In developed countries, this intensified after the Second World War thanks to the use of synthetic nitrogen fertilisers, the development of mechanisation and the use of fungicides, insecticides and weedkillers. The problem is that natural plant populations, or those resulting from domestication, cannot meet all these requirements.

The genetic improvement of plants is therefore aimed at developing populations called varieties which, depending on species and situation, are more productive, more resistant to disease and insects, make better use of water and nitrogen, are better adapted to growing environments or conditions of use (including mechanisation) and have better qualities (nutritional, technological, etc.).

The aim is to combine as many characteristics, and therefore genes, as possible in a single variety, favourable for the objectives sought.

## 2. CREATION

Biodiversity can be conserved, but it can also be created. Varietal creation can be induced by hybridisation or simple mixing between two or more selected varieties. Varietal creation can also be fortuitous, as in the case of natural hybridisation, natural mutations, etc. Plants identified by man in this way can be multiplied and selected to give rise to new varieties. This is the set of processes aimed at bringing together characteristics of interest within a new cultivated plant variety. These characteristics are generally sought in existing varieties or in wild plants of the same species or closely related species (Fig 2)."The species have several cultivated varieties or cultivars. Cultivar or "cultivated variety": artificial population within a species, characterised by:

- A narrow genetic base (even reduced to a genotype);
- Well-defined homogeneous agronomic characteristics ;
- Its reproducibility according to a fixed and registered selection scheme.

### 2.1.Selection criteria :

- Better adaptation to the environment (cold, drought, salinity, etc.)
- Increase productivity (major objective of the green revolution in the second half of the $^{20th}$ century)
- Improving nutritional quality and palatability (biochemical composition, qualities, etc.) organoleptic, appearance...)
- Optimise processing (easier hulling of seeds, better bread-making, etc.)
- Optimise storage and transport and, more recently, protect the environment (pathogen resistance, herbicide tolerance, better use of fertilisers or water, etc.).

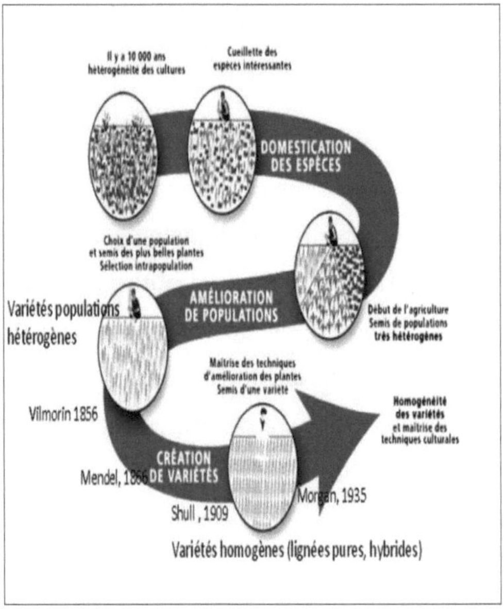


Fig 2: **Stages Plant Improvement**. (gnis - pedagogy)

**2.2.The different types of selection :**

**2.2.1. Massal or visual selection** (until end of the 19th century):

This is the mass selection of a plant population according to phenotypic criteria specific to each farmer who cultivates that population. This ancestral selection technique makes it possible to improve the average value of all the individuals in the population. It consists of retaining certain plants with interesting characteristics in a population, the seeds of which will be used as seed the following year (Fig 3). According to Bouharmont (1994), this method has long been successfully applied to many cross-pollinating species. It involves harvesting seed for the next generation from plants that correspond to the farmer's wishes. Selection is effective for highly heritable traits, and results in a gradual increase in the frequency of favourable alleles in populations; it is slower than in self-pollinated species, especially when several genes are involved. Observation of the phenotype is sufficient for fruit colour, inflorescence morphology or earliness.Mass selection for yield is more difficult: a progeny test involving a few dozen plants is needed to check the value of the individuals identified in the field.

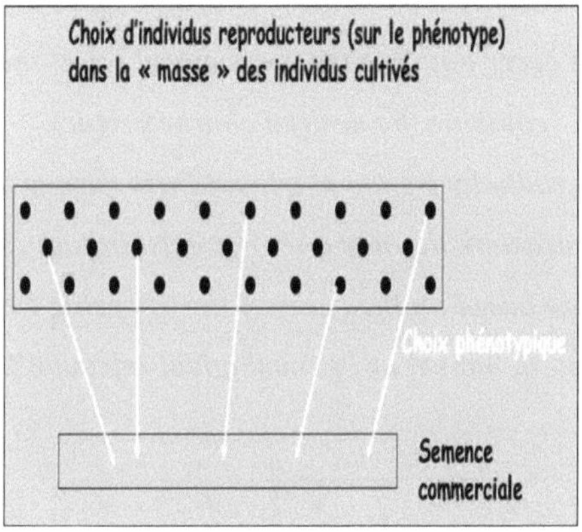


*isual selection of interesting individuals &. Use of their seeds as seeds. The plants selected are different between and from one generation to the next*

***(Population varieties).***

Fig 3. **Diagram of Mass Selection**

**2.2.2. Modern or genealogical selection** (from the 20th century) :

This is a process which consists of deliberately crossing plants which have the following characteristics that we want to perpetuate and/or associate (Fig 4). The discovery of sexual reproduction in plants (1676, Millington-Grew) and the rediscovery of Mendel's laws (1865) by De Vries (1903) laid the scientific foundations for modern plant breeding:

***Plants can be crossed to combine traits of interest.***

***After the first cross, the first generation (F1) is homogeneous, then the characteristics segregate (chromosomal recombination) in subsequent generations.***

***"Some plants have the characteristics of one or other of their parents, while others combine these characteristics".***

***Seed companies carry out directed crosses and select the offspring that combine the desired characteristics.***

***In this way, they gradually isolate a new uniform variety (a combination of characteristics that does not exist in nature).***

***Selection is therefore based on the traits of the offspring rather than the plant itself, known as "genealogical selection".***

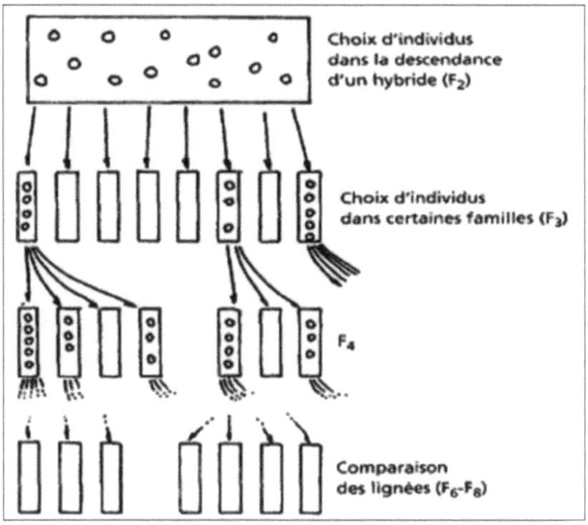


Fig 4. **Schematic diagram of pedigree selection** (Bouharmont, 1994).

## 3. SELECTION AND HYBRIDIZATION

According to Bouharmont (1994), the improver exerts more or less strong selective pressure on heterogeneous populations, by eliminating individuals that deviate from the desired type or choosing a few plants that will produce the next generation (Fig 5).When a trait is highly heritable, selection is generally easy. For other traits, observation of an individual is not enough; its genotypic value must be checked after clonal multiplication or seeding of its progeny.Cross-pollinating populations are sufficiently heterogeneous for selection to be effective. In self-pollinating populations, old varieties are also polymorphic, but modern varieties are very uniform: in this case, the search for new genetic combinations is preceded by crosses between several different genotypes (Fig 6).In general, it's enough to obtain a few hybrid plants: the flowers are castrated, isolated and pollinated by hand. It is often sufficient to combine two parents chosen for their complementary characteristics. However, a population resulting from successive crosses between a larger number of genotypes represents a potential source of exceptional genetic combinations.

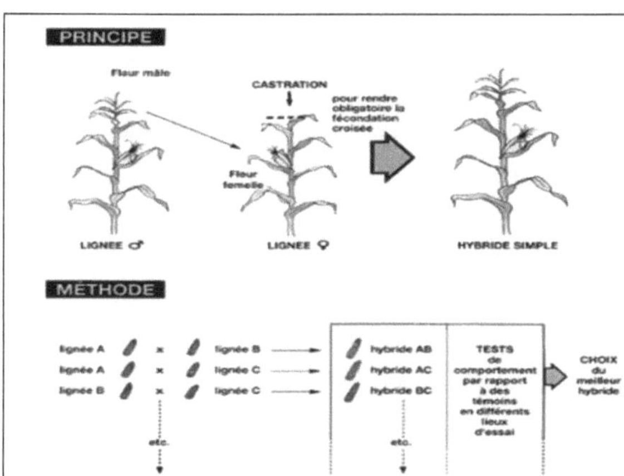


Fig 5. **Steps in making a hybrid** (gnis - pedagogy)

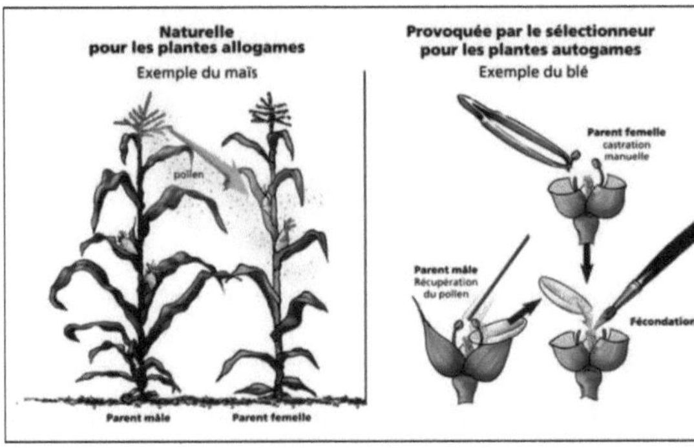


Fig 6. **Example of hybridization in maize and wheat** (gnis - pedagogy)

**3.1.Genetic basis of selection**

**3.1.1.** Inbreeding **depression :**

This is a reduction in plant vigour and fertility linked homozygosity resulting from self-fertilisation (linked in particular to the expression of deleterious recessive genes).According to Bouharmont (1994), in many cross-pollinating plants, self-fertilisation and inbreeding reduce plant vigour, fertility and production to a greater or lesser extent. This depression is significant in maize, where the best inbred lines obtained by self-fertilisation produce half as much as the initial populations. Other species, such as alfalfa and carrots, are even more sensitive to self-fertilisation, which results in a large number of lethal plants.

On the other hand, other species are not very sensitive to inbreeding (rye, sunflower, cucurbits) and it is possible to obtain pure, non-degenerate lines.

Depression can be explained, at least in part, by the existence of numerous recessive lethal alleles (albinism) or disadvantageous alleles, which only appear in the homozygous state, i.e. mainly after self-fertilisation. According to Verrier et al (2001), an individual is 'consanguineous' if his or her parents are related to each other. them. Certain breeding systems lead to systematic appearance of inbreeding.Some plants, known as autogamous, reproduce naturally by self-fertilisation, which rapidly leads to inbreeding of close to 100%. More generally, the structuring of a large population into small sub-populations that reproduce within themselves leads to a regular increase in inbreeding within the population as a whole and differentiation between sub-populations due to genetic drift. This phenomenon is accompanied a deficit in heterozygotes compared with what would be expected.

By modifying the genetic structure of populations, inbreeding have an impact on quantitative traits.

### 3.1.2. **Hybrid vigour** (Heterosis):

The heterosis effect, also known as hybrid vigour, is the increase in hybrid vigour compared with the parents, linked to the heterozygous state. It means that the hybrid individual is superior to the average of the two parents, or to the best of the two parents, in many traits (vigour, yield, disease resistance and earliness). The further apart the parents are genetically, the greater the hybrid vigour. In plant genetics, the term "hybrid" always refers to the result of crossing two different parental populations belonging to the same species. When the union between genitors belonging to two different species is practised, it is specified by using the term inter-specific hybridisation (Verrier et al. 2001). The creation of pure and stable lines, the future parents of the hybrid, is an essential step in the creation of homogeneous and reproducible hybrid varieties expressing the most heterosis. The creation of hybrid varieties therefore involves two phases leading to the selection of the best lines:

- the creation of lines which will serve as good genitors or good parents (from germplasm of superior quality) for the future hybrid by successive self-fertilisation, to fix the characteristics of interest;
- the choice future parental lines by crossing them with a tester to determine their suitability for combination, i.e. to produce a hybrid variety in line with the breeder's objectives.

According to Bouharmont (1994), plant vigour is restored when two lines from self-fertilisation are crossed. Often, vigour is also increased when two heterozygous populations are crossed. This hybrid vigour or heterosis translates into better overall plant health: growth, yield, earliness, tolerance to pests and unfavourable weather conditions. The combination in the hybrid of dominant alleles present in the parents may partly explain the heterosis. Some authors favour this interpretation, believing that most of the heterosis can be fixed in the homozygous state (Galláis, 1988). However, although the many efforts made since the beginning of the century have improved the standard of maize lines, the difference between purified lines and hybrids has not diminished. Although the precise mechanisms responsible for heterosis remain unexplained, superdominance remains the best interpretation. Vigour is attributed to the heterozygosity of hybrids for a large number of genes (Bouharmont, 1994):

- Advantages: Heterosis, phenotypic homogeneity, performance, homeostasis, guaranteed seed quality.
- Disadvantages: Complexity of seed production, impossible to resow from one year to the next, dependence on farmers, high cost of seed, problem of genetic homogeneity for susceptibility to specific pathogens.

#### 3.1.2.1. The different types of hybrid

Crossing two lines produces a variety known as a single hybrid. A single hybrid can be crossed with a selected line to produce a three-way hybrid, or with another single hybrid to produce a two-way hybrid (Fig 7) :

• Single hybrids are trickier to produce because their parents are lineages, but commercial hybrids are homogeneous and have maximum heterosis.

• The three-way hybrid is easier to produce, and it is generally the hybrid that is chosen as the female parent in this case. As a result, seed production is high. It also ensures a broader genetic base for the final variety, by combining traits found on three parents.

• Dual hybrids, which are less efficient than single hybrids, and three-way hybrids are no longer on the market.

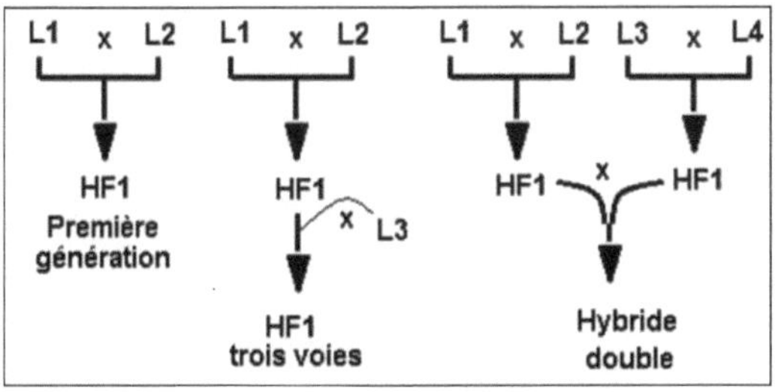


Fig 7. **Different hybrid varieties**

# 4. HOW DO YOU INTRODUCE A NEW TRAIT INTO AN ELITE VARIETY ?

## 4.1 Introgression by hybridisation then backcrossing

An interesting trait, such resistance to disease, tolerance to abiotic stress, sterility or improved quality criteria, may be present in a plant but not in the elite lines on which commercial varieties are based. The breeder will try to create a line that is identical to the elite line, but which also possesses this trait. To obtain this result, the breeder proceeds by back-crossing (Fig 8).

To do this, it carries out a series of hybridisations between the recipient or elite line and the donor line for the trait. The offspring are then crossed for several generations by the recipient or recurrent line. This increases the proportion of the elite line in the genetic background of the offspring, while ensuring that the trait of interest is retained by eliminating undesirable individuals. The result of backcrossing is a converted line, i.e. virtually identical to the elite recipient line, but with the addition of the interesting trait (Fig 8).

In field crops, only hybrids between two cultivated species acceptable agronomic qualities. Even under these conditions, there is a tendency to eliminate most of the characteristics of one of the parents (rye, robusta coffee, Oryza glaberrima). The aim of crosses between cultivated and volunteer varieties is always to transfer a limited number of genes from the latter to the cultivated varieties. The methods available differ according to the level of homology between the parental genomes. They are generally very time-consuming to apply, but once a foreign gene has been introduced into one variety, it can be easily transferred to others by back-crossing. When the chromosomes of two species are able to pair up and form crossovers to carry out genetic recombinations, for example when crossing a cultivated plant with its non-domesticated ancestor, the methods used are not very different from those used when crossing two varieties, except that problems of incompatibility and sterility often arise. This is how many characteristics have been introduced into tomatoes from a few wild species (Bouharmont, 1994).

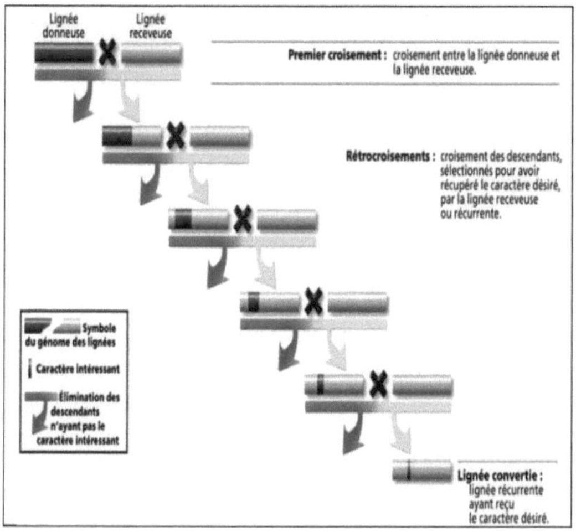


Fig 8. **Introgression principle** (gnis - pedagogy).

### 4.2. Introgression by transgenesis plants

Transgenesis has made a resounding entrance into plant improvement, as it makes it possible to introduce genes that do not exist in the species to be improved. Examples include genes for herbicide tolerance or insecticide molecule synthesis (Gaufichon et al. 2010).

Transgenesis or genetic transformation is the integration of an exogenous DNA molecule into a genome. This molecule may be native or constructed. A genetically modified organism or plant has a very broad biological meaning, since any living being is genetically modified by mutation and recombination in relation to those that gave rise to it (Pelletier, 2012). The first transgenic plant to be cultivated was a delayed ripening tomato (1993).

Genetic engineering refers to all the techniques used to introduce one or more genes from any other and have them expressed in a living organism. Organisms obtained in this way are known as Genetically Modified Organisms (GMOs). A distinction is made between molecular biology techniques, which make it possible to prepare the DNA sequences that will be introduced, known as genetic construction, and transgenesis techniques, which make it possible to transfer the gene. This is a case where gene transfer is accompanied by trait transfer. A copy of the gene of interest is introduced into the plant. Its expression, via messenger RNA, leads to the production of a protein responsible for the new trait.

There are many examples in this field: introduction of a gene for resistance to insects, pathogens or herbicides, modification seed , production of molecules of industrial or pharmaceutical interest.

Stages of transgenesis (Fig 9 **):**

1st step: Identify the gene of interest: Know its sequence and function,

2th step: Make a gene construct,

3(rd) stage: Transformation of the plant and regeneration by in culture,

4th stage: Characterisation of transformants: Selection of the best individuals, transfer and analysis in the greenhouse and in the field.

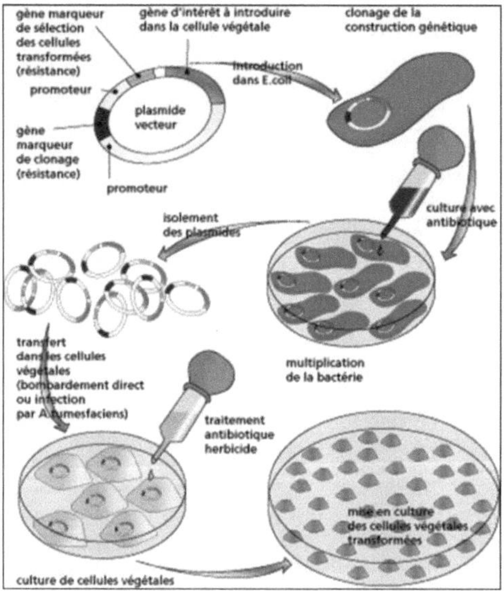


Fig 9. **Stages of plant transgenesis**

## 4.3. Introgression via Agrobacterium tumefaciens

Transgenesis involves adding a new gene to an organism. Several transgenesis techniques have been developed for plants. The possibility of regenerating an entire plant from just a few plant cells is of great interest in transgenesis. One of the most widely used techniques in plant transgenesis is the use of a soil bacterium, Agrobacterium tumefaciens. This gram-negative, rod-shaped bacterium of the Rhizobium family, known as Rhizobium radiobacter (Young, 1889), is attracted to phenolic compounds released by dicotyledonous plants when they are injured and causes a tumour, crown gall, in infected plants (Fig 10; Fig 11).Since ancient times, we have known about a disease known as "crown gall", i.e. the formation of tumours in the area where the root meets the stem, which affects many cultivated plants such as cabbages (Brassica) as a result lesions leading to necrosis of the aerial part of the plant and ultimately its death. This is due to cultivation practices (injury, pruning), climatic factors (frost, hail), pests (nematodes, insects) and natural causes (leaf scars). The disproportionate enlargement of the stems of affected plant species is one of the characteristic symptoms, to plant dieback. Since 1974, it has been known that this induction is due to the transfer of a small plasmid DNA from the bacterium into the genome of the plant's cells. In this wound, Agrobacterium is able to attach itself to the 's cells. Following this contact, these cells multiply significantly, giving rise to a tumour formation. The gall cells release specific chemical compounds into the environment: opines, molecules generally comprising an amino acid linked to a sugar, act as nutrients for the agrobacteria and activate their multiplication, but the synthesis of these products diverts certain pathways of the plant's normal metabolism (Tourte, 2002). The Agrobacterium bacteria present near the gall in the soil are then able to use these opines as a source of nitrogen, as well as carbon and energy (Weidner and Furelaud, 2003). The mechanism by which the galls form is similar to conjugation, and is due to a bacterial plasmid called Ti plasmid, which makes the bacteria virulent (Fig 12).

There are three classes of plasmids depending on the nature of the opine:

• octopine plasmids in which the sugar pyruvate and the amino acid either arginine, ornitine, lysine or histidine ;

• agropin plasmids whose amino acid is glutamine, generally associated with mannose ;

• nopaline plasmids in which the arginine or ornitine is linked acetoglutarate.

The Ti plasmid comprises several regions, three of which are particularly important (Fig 13):

• a region containing the origin of replication ;

• a region corresponding to the transfer DNA (T-DNA), bounded by two borders known as the right and left borders and containing the tumour's oncogenes;

• a vir region containing various genes influencing the bacterium's virulence (ability to recognise plant cells, control the transfer of the T-DNA to the nucleus and its integration into the plant cell genome). These vir genes, located on the plasmid but outside the T-DNA region, are essential for transfer. There are several vir genes - at least six - located linearly on the plasmid, in a region close to the origin of replication (Tourte, 2002) which are described in the following table:

Table 1. **Function of the virulence genes (vir) of the Ti plasmid** (Tourte, 2002)

| Vir genes | Function |
|---|---|
| vir A | Encodes the synthesis of a wound signal receptor, sensitive to the molecules (polyphenyls) released during injury. |
| vir B | Codes the synthesis a membrane protein involved in the formation of the T-DNA transition bacteria to plant cell |
| vir C | Directs the synthesis a protein capable of directing the transfer of T-DNA by positioning the head of the molecule. |
| vir D (D1and D2) | Responsible for synthesis an endonuclease acting at the edges and involved in the release T-DNA. |
| vir E | Encodes the synthesis of proteins protecting T-DNA against nucleases |
| vir G | Responsible for the synthesis of a virulence activating |

A DNA fragment of the Ti plasmid, the T-DNA, is transferred from the bacterium to the plant, then integrated into the plant genome where it induces the formation of galls characterised by the anarchic multiplication of plant cells (Fig 14). This observation forms the basis of one of the most widely used techniques in plant genetic engineering to produce Genetically Modified Organisms or Plants, GMOs or GMPs respectively (Fig 15). In 1983, the first transgenic tobaccos were obtained using agrobacteria simultaneously by Belgian and American teams. Figure 16 summarises the stages in transgenesis using Agrobacterium tumefaciens.

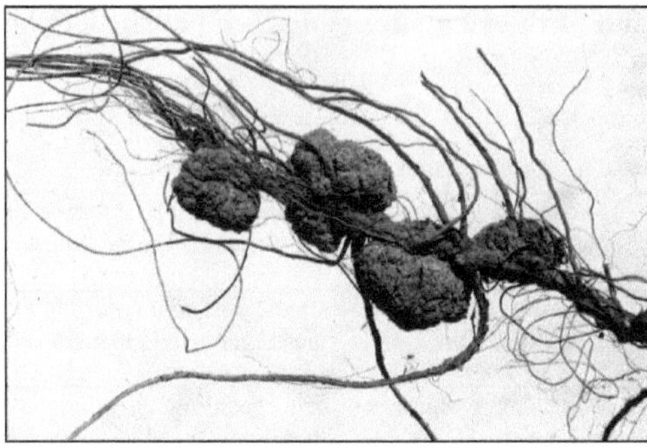

Fig 10. **Crown gall caused by Agrobacterium tumefaciens on a pecan root.**

(https://fr.wikipedia.org/wiki/Agrobacterium_tumefaciens).

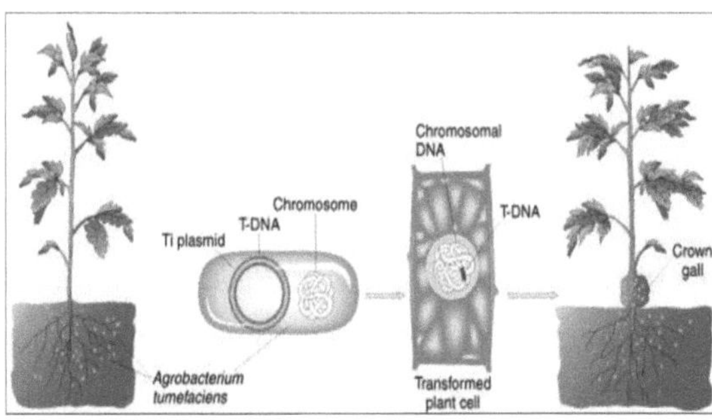


Fig 11: **Infection of the plant by Agrobacterium tumefaciens and formation of crown gall.**

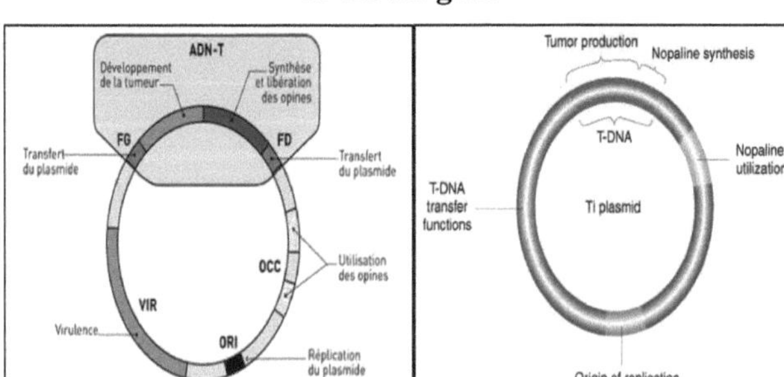


Fig 12. **Structure of the pTi plasmid (215 Kb)** (Weidner and Furelaud, 2003)

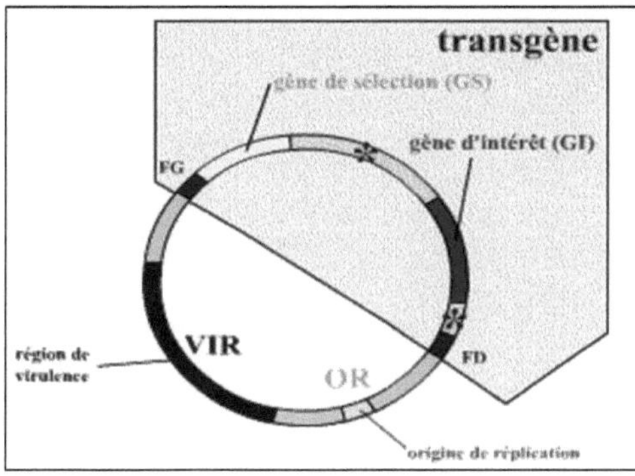


Fig 13. **Plasmid pTi** (disarmed)**: Transgenesis vector** (Weidner and Furelaud, 2003)

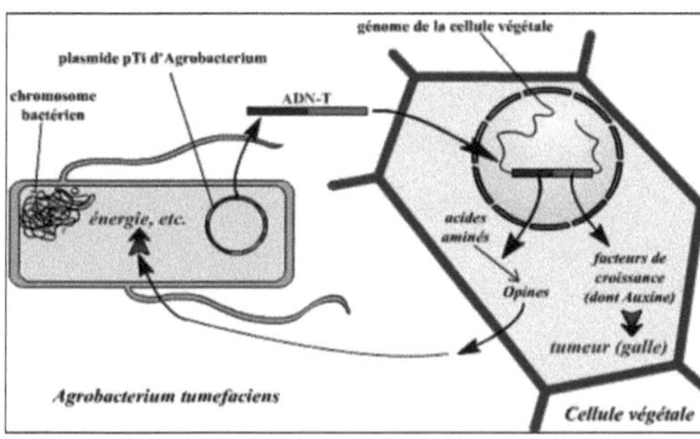


Fig 14. **T-DNA transfer in the plant genome via Agrobacterium tumefaciens** (Weidner and Furelaud, 2003)

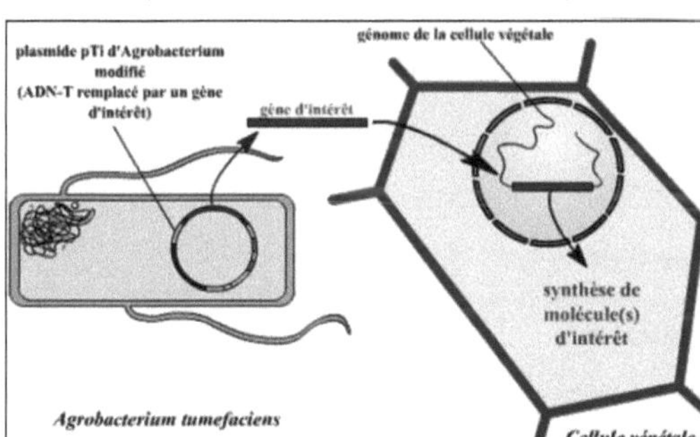


Fig 15. **Replacement of T-DNA with the "PGM" gene of interest**

(Weidner and Furelaud, 2003)

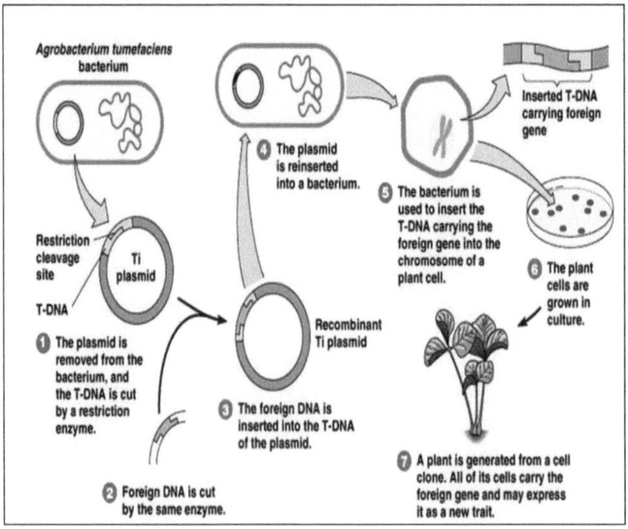


Fig 16. **Summary of the steps involved in transgenesis using Agrobacterium tumefaciens**

(2004, Pearson Education, inc, publishing as Benjamin Cummings)

1. Extraction Plasmid Ti+ DNA cutting - T / Restriction Enzyme. 2. Foreign DNA (Insert) cut / Same Enzyme. 3. Foreign DNA Inserted into -T-DNA Plasmid. 4. Recombinant Plasmid Reinserted into Agrobacterium. 5. Bacterium Used to Insert -T-DNA Carrying Gene of Interest (Foreign) / Plant Cell Chromosomes. 6. Cell Culture.

7. Plant regeneration from clone cells. Cells carrying the gene of interest: expression as a new trait.

**Some Useful Definitions :**

**Agrobacterium :**

Bacteria that parasitise plants by transferring part of their plasmid, called T-DNA, into the genome of the infected plant.

**Allele :**

Part of the chromosome containing the gene or version of a gene that is passed on to offspring.

**Allogamy :**

Cross-pollinating plants are plants that do not self-pollinate or do so only to a very limited extent. The flower's pistil is accessible to pollen from another flower and can therefore be pollinated by the wind or insects. In this case, there may be cross-fertilisation between varieties and the seeds may not reproduce the variety from which they originate: this is known as hybridisation. Some plants are self-compatible, i.e. they can pollinate themselves if no other pollen is available (preferential allogamy, e.g. sunflower). Other so-called self-incompatible plants, such as cabbage, cannot self-pollinate: the female flower of a plant refuses pollen from that same plant, so the plant is necessarily allogamous.

**Self-fertilisation :**

Fertilisation of an ovule by pollen from the same plant.

**Autogamy :**

So-called "self-pollinating" plants pollinate themselves (self-fertilisation), with pollen from one plant fertilising the ovary of the same plant. It is the shape of the flower that is responsible for this protection against cross-pollination between varieties: as the flower is relatively closed, the pistil does not protrude, so pollen from outside cannot penetrate and settle inside. Only pollen from the stamens of the same flower can be deposited on the pistil. As a result, there is a good chance that there will be no cross-breeding between two varieties, even if they are side by side. Nevertheless, we should speak of a predominantly self-pollinating plant, as there are practically no strictly self-pollinating plants: there are always a few percentages of cross-pollination.

**Backcross :**

Crossing a hybrid with one of its parents.

**Biotechnology :**

All the methods and techniques that use living organisms or their components to manufacture or modify products, to improve plants or animals, or to develop micro-organisms for specific applications.

**Host cell :**

A cell harbouring foreign genetic material supplied by a virus, plasmid, DNA recombined in vitro, or a whole cell.

**Clone :**

Cell isolated, maintained in culture and replicated identically for molecular and cellular biology studies.

**Cloning** :

A method of in vitro cell multiplication by asexual reproduction resulting in the formation of genetically identical individuals called clones.

**Conjugation** :

Natural transfer plasmid or chromosomal DNA from one bacterial cell to another via a cytoplasmic bridge. By extension, gene cloning refers to the technique isolating and amplifying DNA fragments in a cell clone.

**Genetic construction :**

DNA sequence intended for transfer into a cell, comprising a gene of interest, the promoter and regulatory sequences essential for its expression and regulation in the recipient cell and a marker gene.

**Chromosome :**

A structure contained within the nucleus of the cells of all living organisms, which can be observed during cell division and carries genetic information. It is a cellular component made up of a coiled DNA molecule associated with proteins called histones.

**Domestication :**

Adapting plants to human needs, adapting wild plants to cultivation.

**Restriction enzyme** (Endonuclease) **:**

A bacterial enzyme that recognises a specific DNA sequence, the restriction site, and cuts the two strands of the DNA molecule within this sequence.

**Species :**

Group of individuals with similar morphological, physiological and chromosomal characteristics who can interbreed.

**F1 :**

All plants resulting from the crossing of two distinct parents

**F2 :**

Set of plants resulting from an F1 plant after 1, 2, n successive forced self-fertilisations.

**Setting characters :**

Action aimed at obtaining stable homozygous material, generally by repeated self-fertilisation from heterozygous material.

**Gene :**

A unit of genetic information occupying a specific position (locus) in the chromosome. A gene is a segment of DNA that includes the sequence coding for a protein, and the sequences that enable and regulate its expression. Genes determine or influence the expression of a living being's phenotype (shape, colour, various aptitudes, etc.). All the genes make up its genome, or hereditary genetic heritage (germplasm).

**Gene of interest :**

The gene responsible for a trait considered to be of interest, which is transferred to another organism.

**Marker gene :**

A gene whose expression enables the cells containing it to be screened, or a gene which confers on a cell or organism a simple property enabling the cells or organisms carrying it to be identified and/or selected.

**Generation :**

A period of time corresponding to the interval between each degree of parentage.

**Genetic engineering** :

A set of techniques for modifying a cell's hereditary heritage by manipulating genes in vitro (gene transfer).

**Genome :**

All the genes, the hereditary heritage contained in every cell of every living organism.

**Genotype :**

Genetic constitution of an individual or set of genetic characteristics of an individual. Its expression leads to the phenotype.

**Heterosis or hybrid vigour :**

Refers to the fact that crossing lines or populations produces offspring with superior performance to those of the parents. The greater the genetic distance between the parents (lines, population), the better the performance.

**Heterozygosity :**

Presence of two different alleles of a gene on homologous chromosomes.

**Homozygous :**

An individual whose cells possess the gene for a given trait in duplicate. This property confers on the individual that possesses it the ability to transmit the trait to 100% of its descendants. A fixed line, or pure line, is totally homozygous. It therefore transmits exactly the same genetic heritage to 100% of its descendants.

**Homozygosity :**

Presence of two identical alleles of a gene on homologous chromosomes.

**Hybrid :**

A hybrid is the crossing of two individuals of two different varieties. However, hybridisation is different genetic manipulation in that it can be caused man, but can also occur naturally by insects or the wind. F1 hybrid This is a 1st generation (1st year) hybrid with a mixture of the genetic characteristics of both parents. When the parents come from pure lines, the hybrid is very homogeneous. This type of hybrid can be found in seed companies' catalogues. F2, F3 hybrid... F2 means that this is the 2nd generation of a hybrid, i.e. a reseeding of individuals from F1 hybrids. This reseeding will produce a set of heterogeneous plants, some resembling the F1, others the "father" and others the "mother".

**Hybridization :**

Cross-fertilisation of the ovule of one plant by pollen from another plant of the same species.

**Insert** :

Sequence of foreign DNA introduced into a given molecule of DNA

**Insertion** :

Additionof a foreign DNA sequence to a given DNA molecule .

**Integration** :

Process of genetic recombination that inserts one DNA molecule into another.

**Marking :**

Introduction of modified nucleotides, or chemical modification of certain nucleotides in a nucleic acid so that it can be identified.

**Pure line :**

A pure line is a set of homozygous individuals or populations all possessing a fixed common trait and able to transmit it indefinitely from generation to generation (e.gindividuals in a pure line of round-fruited tomatoes will give round-fruited progeny by reproducing among themselves).

**Locus :**

Location occupied by a gene on the chromosome.

**Plasmid** :

Extrachromosomal DNA capable of independent replication and carrying genetic characteristics that are not essential to the host cell.

**Recombinant plasmid** :

Plasmid into which a fragment of foreign DNA has been inserted.

**Ti plasmid**:

The main vectors routinely used to produce transgenic plants are derived from the soil bacterium Agrobacterium tumefaciens.

**Population :**

A population is a group of individuals of the same species who reproduce freely among themselves during their lives in a biological environment to which they are adapted.

**Precocity**

This is an expression of the length of a plant's development cycle between sowing and flowering and harvesting at maturity. The earlier the variety, the fewer units of heat to reach flowering or maturity. A very early variety has a short cycle as opposed to a late variety with a long cycle.

**Genetic recombination :**

Exchange between homologous sequences of two chromosomes leading to the appearance in a cell or individual of genes or hereditary characteristics in a different association to that observed in the parental cells or individuals.

**Resistance** :

The origin of resistance is generally genetic and linked to the presence of a resistance gene. Resistance genes may be naturally present in certain varieties, or introduced into a hybrid by backcrossing, or present in other species and introduced by gene transfer, as the case of resistance to the corn borer, a butterfly pest. In the case of resistance, the variety protects itself by reacting against something that destroys it (parasite, disease, etc.). A distinction must be made between resistance and tolerance, although the two terms are often used indiscriminately (see Tolerance).

**Genealogical selection :**

Selection based on obtaining lines or families of lines with the choice of the best of .

**Selector:**

Person in charge of creating new varieties for seed companies. Also known as a breeder.

**Genetic sterility :**

Male sterility determined solely by one or more nuclear genes

**Male sterility :**

Sterility of the male organs of a plant through atrophy of the stamens or more or less early abortion of the pollen in the anthers. This results in the absence of pollen or non-viable pollen.

**Cytoplasmic male sterility :**

Male sterility system involving both the cytoplasm via mitochondrial genes and specific nuclear sterility or fertility restoration genes.

**Tolerance :**

The ability of a variety to withstand the development of a pest or pathogen without the resulting disorders compromising its growth or production: Tolerance to biotic and biotic stresses.

**Totipotence :**

The ability of a cell to regenerate a complete individual identical to the parent plant. It is based on the ability to dedifferentiate: cells can become simple, non-specialised cells and then differentiate to give different types of specialised cells once again.

**Transgenesis** :

All the operations involved in obtaining transgenic organisms.

**Transgenic** :

Describes a living being that is the product of a cell into which foreign DNA has been introduced. The transgenic organism has the introduced foreign DNA in most or all of its cells. The foreign gene can therefore be passed on to offspring.

**Genetic transformation** :

This is the hereditary modification a genome following the integration of a DNA sequence.(transgene).

**Varieties :**

A variety is a group individuals of the same species which share a number of hereditary characteristics.

**Vector :**

Nucleic acid molecule into which fragments of foreign nucleic acid can be inserted and then introduced and maintained in a host cell (e.gplasmid).

# REFERENCES


Gallais A., 2015. Understanding plant breeding: issues, methods, objectives and selection criteria. Quae, Versailles, 240 p.

Bouharmont J., 1994. Agronomie Modèrne. Chapter 13: Varietal creation and plant improvement. Université catholique de Louvain, Louvain-la-Neuve, Belgium

Gaufichon, L., Prioul, J.L., Bachelier, B., 2010. What are the prospects for genetic improvementof drought-tolerant crops? FARM Foundation.

Demarly Y., Sibi M., 1996. Plant improvement and biotechnologies. Montrouge: J. Libbey Eurotext , 1996. 2$^{th}$ ed. Collection Universités francophones.

Verrier E., Brabant Ph., Gallais A., 2001. Institut National Agronomique Paris-Grignon. Georges P., 2012. French Academy of Agriculture. Session of 5 December.

Young J M, L D Kuykendall, A Kerr and E Martínez-Romero, 2001. "A revision of Rhizobium Frank 1889, with an emended description of the genus, and the inclusion of all species of Agrobacterium Conn 1942 and Allorhizobium undicola de Lajudie et al. 1998 as new combinations: Rhizobium radiobacter, R. rhizogenes, R. rubi, R. undicola and R. vitis.", International Journal of Systematic and Evolutionary Microbiology, vol. 51, 1 January, pp. 89-103 (DOI 10.1099/00207713-51-1-89.

Smith EF, Townsend CO. 1907. "A Plant -Tumor of Bacterial Origin", Science, vol. 25, no. 643, April, pp. 671-3 (PMID 17746161, DOI 10.1126/science.25.643.671)

Weidner M, Furelaud G., 2003. Transgenesis using Agrobacterium tumefaciens, Planet-Vie.

Tourte,Y., 2002. Génie génétique et biotechnologies: concepts, méthodes et applications agronomiques. Edt. Dunod, Paris, 241pp.

**Find out more!**

Link to websites :

http://www.gnis-pedagogie.org

https://fr.wikipedia.org/wiki/Agrobacterium_tumefaciens

https://planet-vie.ens.fr/article/1483/transgenese-grace-agrobacterium-tumefaciens

http://www.snv.jussieu.fr/vie/dossiers/transgenese/agrobacterium/agro.htm

# APPENDIXES

## APPENDIX 1

### Additional documents AGP

Document 1

**"More generally, from domestication plant improvement, we are witnessing genetic erosion".**

(Accelerated by the green revolution).

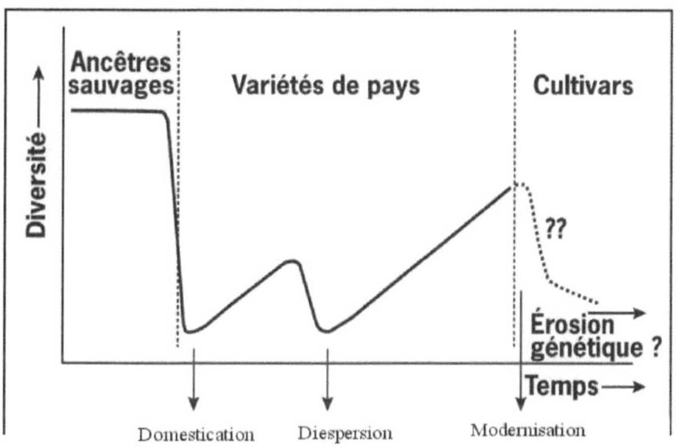


"Modèle de variations temporelles de diversité génétique pour une espèce de grande culture depuis les ancêtres sauvages jusqu'aux cultivars modernes (van de Wouw et al, Genetic erosion in crops, 2009)".

## Document 2

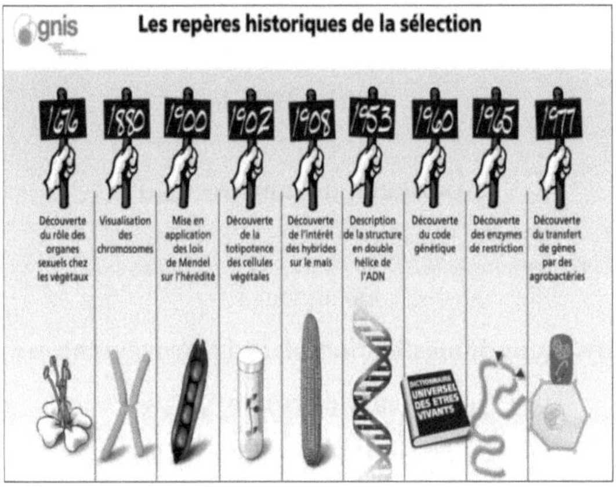


## Document 3

**Plantes autogames. Schémas d'amélioration génétique**

(Grignac et al., 1978)

| Pedigree<br>Sélection généalogique | SSD<br>Sélection 'Single Seed Descent' | Bulk<br>Sélection généalogique différée |
|---|---|---|
| 1. Choix de 300 à 500 Épis dans F2 | 1. Prélèvement de 1 à 2 grains sur chaque plante F2 | 1. Récolte de la totalité des grains. Sélection des 2000 plus gros |
| 2. Famille F3 (300 à 500) Choix sur résistance aux maladies et teneur en protéine des grains | 2. Population F3 Prélèvement de 1 à 2 grains sur chaque plante | 2. Population F3 Récolte de la totalité des grains. Sélection des 2000 plus gros |
| 3. Famille F4 (20 à 50) choix sur résistance, rendement teneur en protéine | 3. Population F4 Prélèvement de 1 à 2 grains sur chaque plante | 3. Population F4 Récolte de la totalité des grains. Sélection des 2000 plus gros |
| 4. Famille F5 (10 à 30) choix comme F4 | 4. Population F5 choix de 500 à 900 épis | 4. Population F5 choix de 500 à 900 épis |
| 5. Famille F6 (6 à 15) choix comme F4 | 5. Famille F6 (500 à 900) choix résistance maladie, teneur en protéine | 5. Famille F6 (500 à 900) choix comme SSD |
| 6. Familles F7 (5 à 12) choix comme F4 | 6. Familles F7 (20 à40) choix résistance maladie, teneur protéine, rendement | 6. Familles F7 (20 à 30) idemF6 |
| 7. Famille F8 (4 à 8) | 7. Famille F8 (10 à 20) | 7. Famille F8 (10 à 20) |

Document 4

**Delayed pedigree selection** (or Bulk method)

Stages :

1. hybridisation

2. No selection in F2

3. Fixation without selection (self-fertilisation a random plant in each generation F3 self-fertilisation)

to F5)

5. selection of the best lines after fixation (F6 to F8)

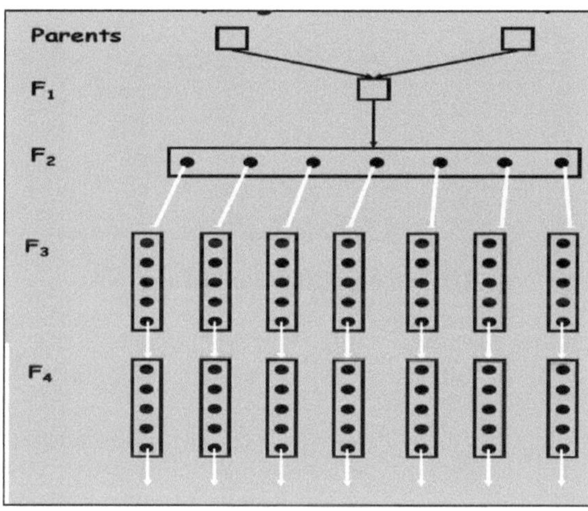

# APPENDIX 2

## Basic elements of Molecular Biology and Genetic Engineering

### a) How do you amplify a gene of interest?

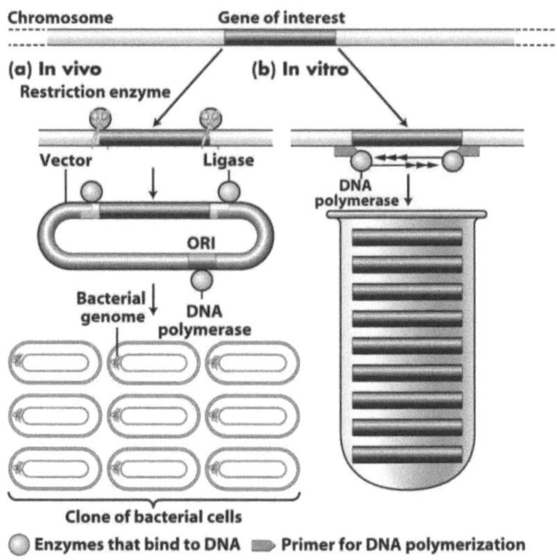


### b) In vivo amplification using DNA

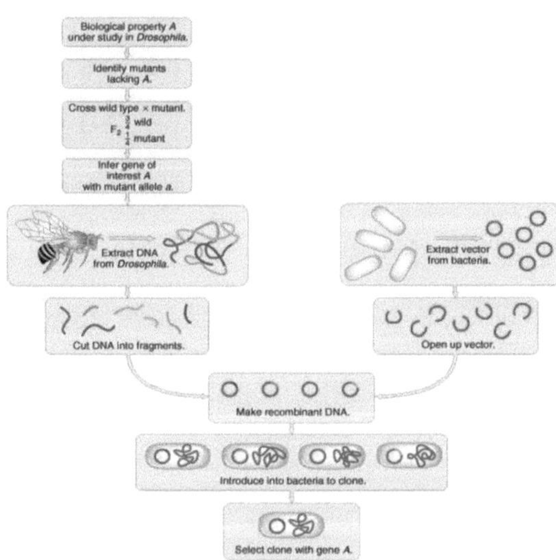

"The combination of a vector and one or more inserted fragments is called recombinant DNA. The recombinant DNA mixture is then used to transform bacterial cells. In general, a single molecule of recombinant vector is inserted into each bacterial cell. These cells are spread out and multiply by forming colonies. As a result, a colony contains a very large population. of identical DNA inserts, and this population is called a DNA clone".

**c) The construction recombinant DNA**

1- Types donor DNA:

a) Genomic DNA

b) Complementary DNA (cDNA)

c) DNA obtained by chemical synthesis

2- Cut donor DNA and vector using restriction enzymes

3- Linking DNA (donor-vector DNA)

4- Amplify each recombinant DNA using the bacterial machinery

"It was the discovery and characterisation of restriction enzymes that led to the development recombinant DNA technology. Restriction enzymes are produced by bacteria and act like scissors, cutting DNA at specific target sequences".

Example of a cut EcoRI:

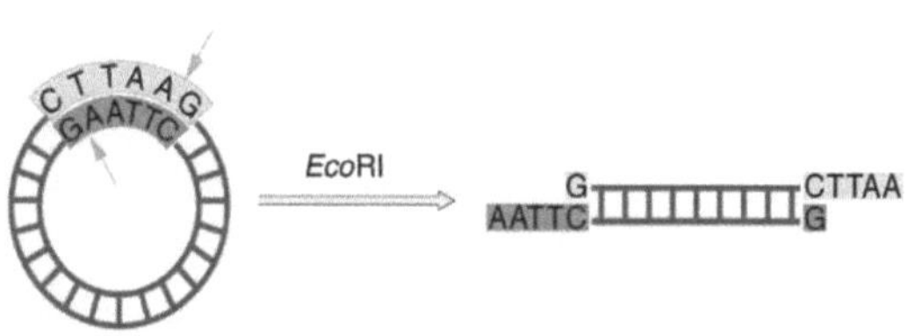

## d) Restriction enzymes in genetic engineering

**Some restriction enzymes**

| Enzyme | Source organism | Restriction recognition site in double-stranded DNA | Structure of the cleaved products |
|---|---|---|---|
| **(a)** | | | |
| *EcoRI* | *Escherichia coli* | 5′ –G–A–A–T–T–C–<br>–C–T–T–A–A–G– 5′ | –G    5′ A–A–T–T–C–<br>–C–T–T–A–A 5′    G–<br>5′ overhang |
| *PstI* | *Providencia stuartii* | 5′ –C–T–G–C–A–G–<br>–G–A–C–G–T–C– 5′ | –C–T–G–C–A 3′    G–<br>–G    3′ A–C–G–T–C–<br>3′ overhang |
| *SmaI* | *Serratia marcescens* | 5′ –C–C–C–G–G–G–<br>–G–G–G–C–C–C– 5′ | –C–C–C    G–G–G–<br>–G–G–G    C–C–C–<br>Blunt ends |
| **(b)** | | | |
| *HaeIII* | *Haemophilus aegyptius* | 5′ –G–G–C–C–<br>–C–C–G–G– 5′ | –G–G    5′ C–C–<br>–C–C 5′    G–G–<br>Blunt ends |
| *HpaII* | *Haemophilus parainfluenzae* | 5′ –C–C–G–G–<br>–G–G–C–C– 5′ | –C    C–G–G–<br>–G–G–C 5′    C–<br>5′ overhang |

## e) Production a chimeric DNA plasmid

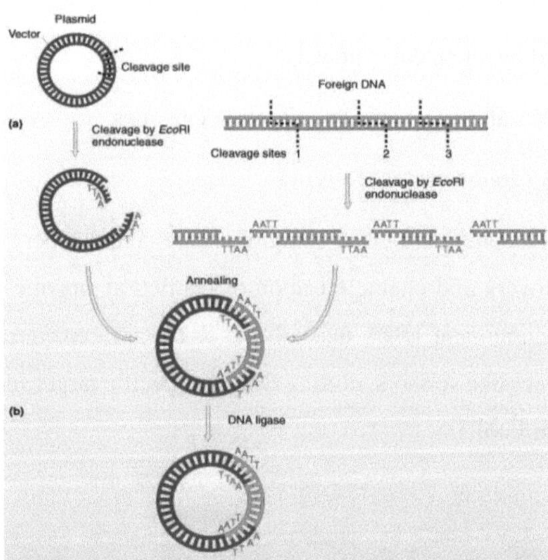

## f) Amplifying recombinant DNA

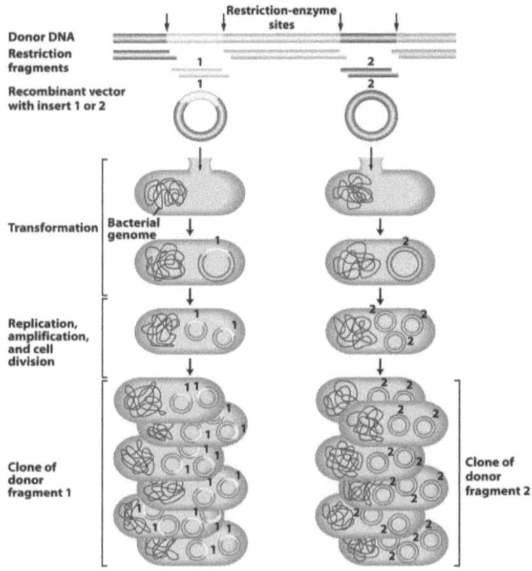


## g) Choosing a cloning vector

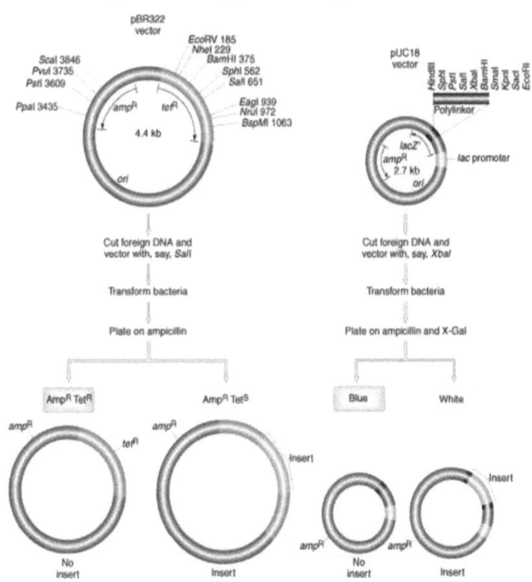

MIX
Papier aus verantwortungsvollen Quellen
Paper from responsible sources
FSC® C105338

Printed by Books on Demand GmbH, Norderstedt / Germany